ANIMAL BABIES

BABY ANTEATERS

by Jennifer Boothroyd

Consultant: Beth Gambro
Reading Specialist, Yorkville, Illinois

BEARPORT PUBLISHING

Minneapolis, Minnesota

Teaching Tips

Before Reading

- Look at the cover of the book. Discuss the picture and the title.
- Ask readers to brainstorm a list of what they already know about anteaters. What can they expect to see in the book?
- Go on a picture walk, looking through the pictures to discuss vocabulary and make predictions about the text.

During Reading

- Read for purpose. Encourage readers to think about anteaters and the animal life cycle as they are reading.
- If readers encounter an unknown word, ask them to look at the sounds in the word. Then, ask them to look at the rest of the page. Are there any clues to help them understand?

After Reading

- Encourage readers to pick a buddy and reread the book together.
- Ask readers to name three things anteaters do between the time they are born and the time they are ready to have babies. Go back and find the pages that tell about these things.
- Ask readers to write or draw something that they learned about anteaters.

Credits:
Cover, © belizar/Shutterstock; 3, © Eric Isselee/Shutterstock; 5, © Leonardo Mercon/Shutterstock; 6, © Tatyana Vyc/Shutterstock; 6, © Eric Isselee/Shutterstock; 7, © belizar73/iStock; 8,9, © Daniel Hernanz Ramos/Getty; 10,11, © mirecca/iStock; 12, © tanukiphoto/iStock; 14,15, © Musat/iStock; 16, © Madlen/Shutterstock; 16,17, © Ondrej Prosicky/Shutterstock; 18, © apple2499/Shutterstock; 20,21, © Luciano Queiroz/Alamy; 22, © mirecca/iStock; 23, © giasch/iStock; 23, © Natalia Bubochkina/iStock; 23, © belizar/Shutterstock; 23, © belizar/Shutterstock; 23, © David Havel/Alamy.

Library of Congress Cataloging-in-Publication Data

Names: Boothroyd, Jennifer, 1972- author.
Title: Baby anteaters / by Jennifer Boothroyd.
Description: Bearcub books. | Minneapolis, Minnesota : Bearport Publishing
Company, [2022] | Series: Animal babies | Includes bibliographical
references and index.
Identifiers: LCCN 2021026682 (print) | LCCN 2021026683 (ebook) | ISBN
9781636913537 (library binding) | ISBN 9781636913605 (paperback) | ISBN
9781636913674 (ebook)
Subjects: LCSH: Myrmecophagidae--Infancy--Juvenile literature.
Classification: LCC QL737.E24 B66 2022 (print) | LCC QL737.E24 (ebook) |
DDC 599.3/141392--dc23
LC record available at https://lccn.loc.gov/2021026682
LC ebook record available at https://lccn.loc.gov/2021026683

For more information, write to Bearport Publishing, 5357 Penn Avenue South, Minneapolis, MN 55419. Printed in the United States of America.

Contents

It's a Baby Anteater!

What is on the back of that anteater?

It is an anteater **pup**!

The baby holds on to its mother.

The mother keeps her little baby safe.

The **furry** pup is no bigger than a toaster.

It has a long nose and tail.

The pup mostly stays on its mother's back.

Sometimes, it crawls down to drink milk from her.

The baby grows bigger every day.

After one month, it spends less time on its mother.

The pup walks around.

It may climb a tree, too.

The pup sleeps a lot.

It covers itself with its tail.

Some anteaters sleep on the ground.

Others sleep in trees.

When it is three months old, the pup starts eating ants.

It uses its claws to dig at an **anthill**.

The little anteater puts its long **tongue** in the anthill.

Its sticky tongue gets covered in ants.

Then, the pup **swallows** the ants.

Yum!

After a few more months, the pup is even bigger.

It is too big to ride on its mother.

But it still stays close.

The anteater is fully grown in one to two years.

Then, it goes to live by itself.

Soon, the anteater will have its own pup.

The Baby's Body

Glossary

anthill a pile of dirt where ants live

furry covered with hair

pup a baby anteater

swallows moves something from the mouth down the throat

tongue a part of the mouth some animals use to eat

Index

Read More

Gillespie, Katie. *I Am a Giant Anteater (I Am).* New York: AV2 by Weigl, 2018.

Meister, Cari. *Do You Really Want to Meet an Anteater? (Do You Really Want to Meet . . . ?).* Mankato, MN: Amicus, 2019.

Learn More Online

1. Go to **www.factsurfer.com** or scan the QR code below.
2. Enter "**Baby Anteaters**" into the search box.
3. Click on the cover of this book to see a list of websites.

About the Author

Jennifer Boothroyd once ate an ant, but it was covered in chocolate.